$$x^n + y^n = z^n$$

... have I really solved Fermat's Last Theorem?

Salvatore Giuliano Franco

Title | … have I really solved Fermat's Last Theorem?
Author | Salvatore Giuliano Franco
Cover image | edited by the author
ISBN | 978-88-91181-95-4

Youcanprint Self-Publishing
Via Roma, 73 - 73039 Tricase (LE) - Italy
www.youcanprint.it
info@youcanprint.it
Facebook: facebook.com/youcanprint.it
Twitter: twitter.com/youcanprintit

Amarcord… would the unforgettable and unforgotten Federico Fellini say … similarly I, mindful of a school of merrier times as memories often are, remember the teacher who explained us how the exposition of a Topic requires a Title, a Development and a Conclusion.

The Title must be clear and contain in a nutshell the thesis which will be argued later.

The Introduction, explanatory and not lengthy, must however capture the interest of the reader or listener.

The Development is the very heart of the Topic and through it also the heart of those already involved and about to live the same adventure of the narrator should beat.

The Conclusion will ensure that going away from the Topic creates some subtle link as that light nostalgia only beautiful dreams can leave.

Hence it is clear, unknown reader friend, that now we find ourselves in the very middle of the Introduction, since the Title of the Topic has already been explained.

I have just begun to go through the first hours of my eightieth year, but, as already in the fifties, I am firmly convinced that even the older age is not only memory of the strength and the beauty of the youth, but is waiting and expectation of what is yet to happen, within us and around us, and that even a second of life contains an endless time.

When I was fifty I wanted to prove to myself that the curve of intelligence does not decrease after the first twenty-five years, thus I underwent the exams of Mensa Italia - The High I.Q. Society: I joined it, then became Secretary of the Lazio Region, under the Presidency of the loved Menotti Cossu, achieving the 155+ score, on the scale which estimated the I.Q. until 148 at the time.

And today what do I still want to prove?

That every type of Search is youth of spirit, love of learning, life.

Handicraft, painting, sculpture, physics, chemistry, mathematics, every commitment of ours in the group of the Liberal Arts and not, is music, is poetry.

And if life is dream, any dream is life.

And I, dear companions of adventure, want to live a dream with you: the solution to Fermat's Last Theorem.

And here we are entering in the Development.

I wanted to reason drawing from the Pythagorean modus operandi, but with that acquaintance of mathematics Fermat could have in the seventeenth century.

The solution given in 1995 by that genius of mathematics responding to the name of Andrew Wiles, argued in one hundred and thirty pages of a mathematics really understandable by the few, does not seem in my opinion compatible with Fermat's affirmations.

In the margin of a text from Diophantus' *Arithmetica* Fermat wrote referring to the expression $x^n + y^n = z^n$, "I found a really wonderful demonstration, but the margin of the book is too small to put it."

Fermat's demonstration would undoubtedly confirm that this equation does not have solutions in integers for $n \geq 3$

And now I ask myself, why has he written "wonderful demonstration"?

Because, I suppose, the solution is as logically linear and comprehensible as the demonstration Pythagoras had given of his same theorem which was earlier only a mathematical conjecture.

Even only to allude to Andrew Wiles' demonstration, I must necessarily refer to a mathematics evolved to levels which are really understandable by the few.

Fermat's equation had been transformed by Frey in an elliptic equation, therefore Taniyama-Shimura's modular conjecture, which asserted that an elliptic equation must be modular and that shapes are classified as the fifth after the four fundamental mathematical operations, created a bridge between much various areas, as Langlands dreamt of, and made the connection to Fermat's equation possible.

A delicious argumentation of Frey carried to the conclusion that Fermat's Last Theorem would have been true if, and only if, Taniyama-Shimura's conjecture had been demonstrated.

Using also Kolyvagin-Flach's method and other mathematical paths conceived, constructed and used by himself, after years and years of incorporeal studies Andrew Wiles managed to turn what was only Taniyama-Shimura's conjecture into a theorem, thus to render Fermat's conjecture a demonstrated theorem.

Using some really obscure mathematics for me, this hint to Wiles' demonstration is useful to better understand the distance between his and Fermat's one.

Therefore, returning to the topic, operating like a skilful bricklayer who, before being all set to work, ensures that the bricks on hand are sufficient for the erection of the planned construction, we, unknown friends and companions of adventure, must ensure the quality and the amount of our bricks, too.

Nevertheless our bricks will have to be similar to those Fermat had in the seventeenth century.

These bricks are **the Numbers**.

Hindu-Arabic: numbers replacing Greek, Roman and other ones.

Rational: integers, fractions, periods, positive, negative, all possibly placed on the axis of the abscissas, on the left and on the right of the initial point "0".

Irrational: numbers expressed in decimal form which continue indefinitely, without a regular or coherent structure, named *algebraic* if of the type " $\sqrt{2}$ " and *transcendental* if of the type "e, π ".

Real: rational and irrational.

Imaginary: the square roots of a negative number, $\sqrt{-1} = i$, possibly placed on the axis of the ordinates starting from the initial point "0", upward the positive ones and downward the negative ones.

Complex: the sum-difference of real and imaginary numbers, possibly placed in the plan given by the four quadrants defined by the horizontal axis of the abscissas and by the vertical one of the ordinates, intercrossing in point "0".

As a matter of fact many other types of Numbers exist, defined by the fathers of mathematics since the earliest times, but all comprised in the group of the Real Numbers, and now, using only these bricks, I am about to get to the heart of the dream, my beloved friends.

We take a cue from the famous

Pythagorean Theorem: $x^2 + y^2 = z^2$

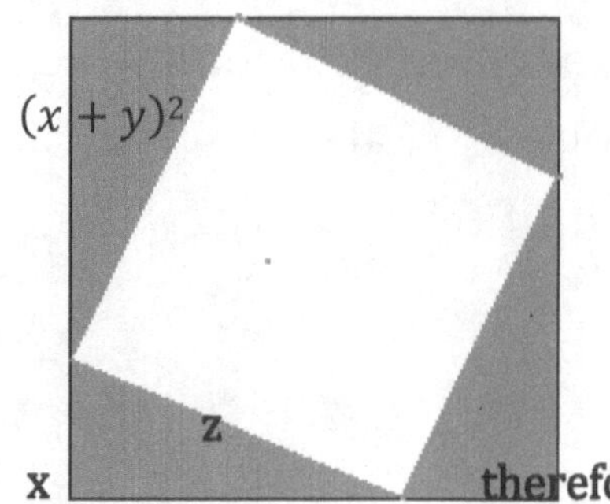

the total area is: $A_1 =$

and, as sum of the parts:

$$A_2 = 4\,\frac{xy}{2} + z^2$$

therefore: $A_1 = A_2$

if: $A_1 = A_2$ also: $(x + y)^2 = 4\,\frac{xy}{2} + z^2$

and, resolving: $x^2 + 2xy + y^2 = 2xy + z^2$

after simplification the initial equation is obtained:

$$x^2 + y^2 = z^2$$

valid for any value of the variables x, y expressed in

rational integers.

Fermat's Last Theorem states that for $n \geq 3$

the equation $x^n + y^n = z^n$ **does not admit solutions**
for **n** rational integer.

Therefore, assuming that the equality is true for $n = 3$

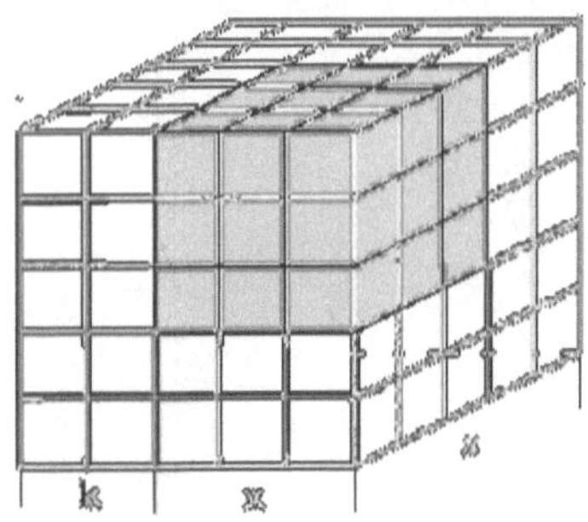

the volume of the great cube is:

$$V = z^3$$

the volume of the small cube is:

$$V_1 = x^3$$

the difference volume be V_2:

$$V_2 = V - V_1$$

Therefore, V_2 is given by the sum of the defined partial volumes and comprised between the great cube and the small cube:

$$V_2 = z^2\,k + z\,x\,k + x^2\,k = k\,(z^2 + z\,x + x^2)$$

$$\text{if} \quad x^3 + y^3 = z^3 \quad \text{equals} \quad V_1 + V_2 = V$$

also $\quad V_2 = V - V_1$ equals $\quad y^3 = z^3 - x^3$

but $\quad V_2 = y^3 \quad$ therefore $\quad y^3 = k\,(z^2 + zx + x^2)$

Is the monomial y^3

really the cube of a rational integer?

We verify it:

For $\mathbf{k = 0}$ $\quad$ also $\quad y^3 = 0$ thus generating the identity

$$x^3 = z^3$$

For $\mathbf{k \neq 0}$ $\quad$ the expression $(z^2 + zx + x^2)$

is resolved in z, at less than a factor k,

$$z = \frac{-x \pm \sqrt{x^2 - 4\,x^2}}{2} = \frac{-x \pm x\sqrt{1-4}}{2} = -\frac{x}{2} \pm \frac{x}{2}\sqrt{-3}$$

but $\sqrt{-3}$ is an imaginary number $= i\sqrt{3}$

therefore $\mathbf{z}$ offers **two** solutions :

$$-\frac{x}{2} + \frac{x}{2}\,i\sqrt{3} \quad \text{and} \quad -\frac{x}{2} - \frac{x}{2}\,i\sqrt{3}$$

each is sum of a real number and an imaginary one, so each solution is a complex number

since $\quad y^3 = k\,(z^2 + zx + x^2)$

the third power of **y**, at less than a factor **k,** is a complex number, external to the group of real numbers and even more external to the set of rational integers, therefore, for **n**

$= 3$ the equation $x^3 + y^3 = z^3$ admits solutions only for a y^3 not belonging to the set of real numbers.

Now we verify the equation $x^n + y^n = z^n$ for $n = 4$

But the equation $x^4 + y^4 = z^4$ equals $x^3 \cdot x + y^3 \cdot y = z^3 \cdot z$

Since "y^3" does not admit solutions if not in the field of complex numbers, also "$y^3 \cdot y$" does not admit solutions in the field of real numbers and even more in the set of rational integers, so also the equation $x^4 + y^4 = z^4$ admits solutions only for a y^4 not belonging to the set of real numbers.

But every value of m elevated to n, "m^n", is equal to the value of $m^3 \cdot m \cdot m \cdot m \ldots$ for a number $n - 3$ of times, thus also the equation $x^n + y^n = z^n$ admits solutions only for a y^n not belonging to the set of real numbers.

Therefore, for any $n \geq 3$ Fermat's Last Theorem is so well verified.

It is very likely, beloved friends and companions of adventure, that you may expect, in more pages, a mathematics even not simple but always very comprehensible, about which you rack your brains as about the solution of a difficult Sudoku.

Nevertheless I think, after elaborating a possibly too simple intuition for a long time, I have centered the heart of the matter, reducing also its demonstration in a very restrained space, even though bigger than the too small margin of the page of a text from Diophantus' *Arithmetica*, as Fermat wrote.

I expect from you a continuation to this short essay of mine.

I wish I would receive possible observations of yours and it would be for me a very mental pleasure to give a continuation to these little pages reporting your words and your name, if desired with some personal news, in a new volume I would give the prints as soon as I should have received a sufficient number of your considerations.

You can address your thoughts to:
sagifra@hotmail.com
and if you were interested in other jobs of mine, not mathematical ones, you can soon find them for the Publishing "youcanprint", on every Italian bookstore and every e-book site, under the name of "Salvatore Giuliano Franco".
And now let me proceed in the conclusion exiting the sown for a while.

I have named you friends and companions, but I would love to clear the term comrade: certainly I am not a Fascist, word currently corresponding to intolerant, absolutist, overpowering, violent; the poet and the mathematician can never be so, anyone really capable of Love will never be despotic.

The word comrade has other origins and very various traditions, it is sufficient to remember that once misused to the times of the two decades, in USSR, from '22 to '53, one used to address to the head of the nation with the name of "comrade Stalin", and not "compagno Stalin", as contrarily translated into Italian.

It is a word of ancient use in the Italian language and is a name which indicates togetherness, sharing the daily struggles for life and the times of rest.

"Comrades" have been the six hundred thousand fallen in the First World War, as well as those of Porta San Paolo, or the 193 officials and the 6500 soldiers slaughtered in Cefalonia, or even those who in prisons, guilty or not, share the little serene moments and the suffering from the lack of freedom, and many others, too many.

The spoils of "comrades" are guarded by the candid urns of the great memorial monument of Redipuglia, which seems to rise to the sky, and by the oldest Ossuary behind it, but also by the white graves of the military cemetery of Mignano di Montelungo, built by my uncle, prof. Arduino Albanese. Or by the nameless sepulcher, one for all, of the Altare della Patria, in Rome.

Words often contain the strength of truth, and truth is always feared, since it rips the mask and reveals the face and the heart, and "comrade" means companion in the right and the duty, in joy and in suffering, but it means also to be able to give oneself for the salvation of another, and this is charity, respect, love.

But anybody on this Earth should feel and be comrade, and Death would be my friend if I had the certainty of being remembered, even only by one person, as "comrade", which is more than friend, companion, partner, brother.

And now farewell, unknown friends, companions of adventure, comrades in the depth of feeling, and do not forget to write to me.

this is not the   **END**